ASTEROIDS

KATE RIGGS

Published by Creative Education and Creative Paperbacks
P.O. Box 227, Mankato, Minnesota 56002
Creative Education and Creative Paperbacks are imprints of
The Creative Company
www.thecreativecompany.us

Design and production by Chelsey Luther
Printed in the United States of America

Photographs by Corbis (Corbis, NASA/Roger Ressmeyer, Detlev
van Ravenswaay/Science Photo Library, Denis Scott, Victor Habbick
Visions/Science Photo Library), deviantART (AlmightyHighElf),
Dreamstime (Mack2happy, Oriontrail), Getty Images (NASA/
Newsmakers, Detlev van Ravenswaay, Stocktrek Images), iStockphoto
(Pingebat), NASA (NASA/ESA/Hubble SM4 ERO Team, NASA/JPL-
Caltech/STScI), Shutterstock (Radoslaw Lecyk, Rainer Lesniewski)

Library of Congress Cataloging-in-Publication Data
Riggs, Kate.
Asteroids / Kate Riggs.
p. cm. — (Across the universe)
Summary: A young scientist's guide to rocky asteroids, including
how they interact with other elements in the universe and
emphasizing how questions and observations can lead to discovery.
Includes index.
ISBN 978-1-60818-480-4 (hardcover)
ISBN 978-1-62832-080-0 (pbk)
1. Asteroids—Juvenile literature. I. Title.
QB651.R54 2015
523.44—dc23   2014002081

CCSS: RI.1.1, 2, 3, 4, 5, 6, 7; RI.2.1, 2, 3, 5, 6, 7, 10;
RI.3.1, 3, 5, 7, 8; RF.2.3, 4; RF.3.3

First Edition
9 8 7 6 5 4 3 2 1

Pages 20–21 "Astronomy at Home"
activity instructions adapted from
NASA's The Space Place:
http://spaceplace.nasa.gov
/asteroid-potatoes/

TABLE OF CONTENTS

Space Rocks	4
Belted or Burning	6
Picture This	8
Ceres	12
Eros and Gaspra	14
Ida	16
What Do You Know?	19
Astronomy at Home	20
Glossary	22
Read More	23
Websites	23
Index	24

Did you know that asteroids are big rocks in space? Scientists called astronomers study asteroids. Most asteroids **orbit** the sun in the asteroid belt. This belt is between the **planets** Mars and Jupiter.

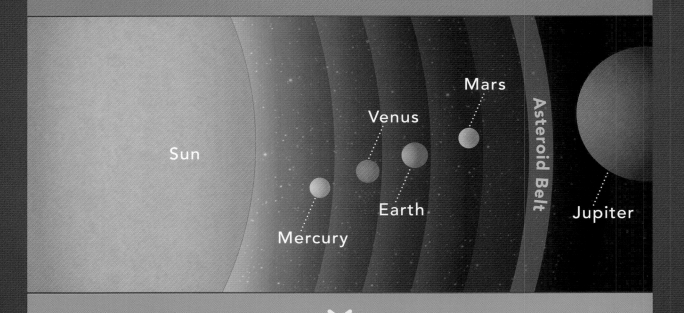

Asteroids look like big potatoes, with pits, or craters, on the surface.

Gravity keeps asteroids inside the belt. Sometimes asteroids run into each other. Or they fall toward something, like a planet. Many asteroids that fall toward Earth burn up before they hit the ground. A burning asteroid looks like a shooting star.

An asteroid could be the reason why all the dinosaurs died out.

Astronomers know about asteroids from **telescopes** and spacecraft. Special vehicles in outer space can take pictures of asteroids. Then astronomers can look at the pictures.

Telescopes at Mauna Kea, Hawaii, USA

Galileo

The Galileo spacecraft was the first to get close-up pictures of asteroids.

Asteroids that are close to Earth are called near-Earth objects.

An astronomer from Italy named Giuseppe Piazzi found Ceres in 1801.

The first asteroid ever found is also the biggest. It is named Ceres. It is about 592 miles (953 km) across. That is as far as if you lived in Omaha, Nebraska, and your friend lived in Indianapolis, Indiana. Ceres is called a dwarf planet. Dwarf planets are things in space that are too small to be planets. But they are not **moons**, either.

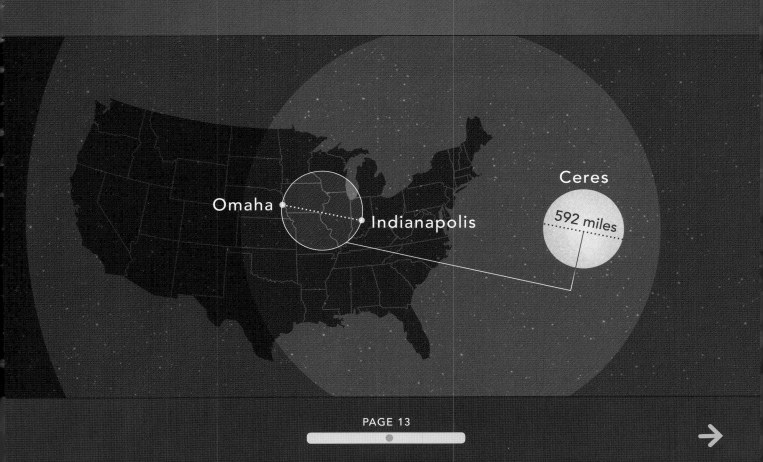

Omaha

Indianapolis

Ceres

592 miles

There are three kinds of
asteroids: C, S, and M.
Eros is an S-type asteroid.

>

Eros is an asteroid that is not in the asteroid belt. It is closer to Earth. Eros looks like a big bone. It is twice as big as New York's Manhattan Island. Gaspra is an old asteroid. Scientists think it has been around for 200 million years!

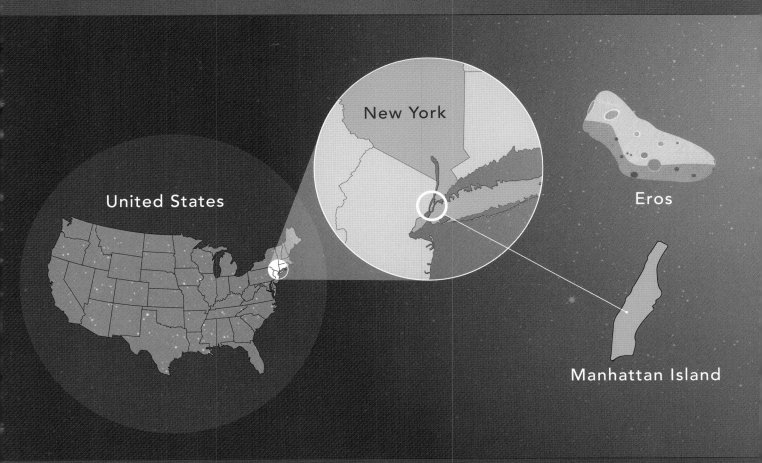

New York

United States

Eros

Manhattan Island

Ida

Dactyl

Asteroids can have moons. The asteroid Ida has a mile-wide (1.6 km) moon named Dactyl. Two asteroids that orbit each other are called **binary** asteroids. These asteroids are usually about the same size.

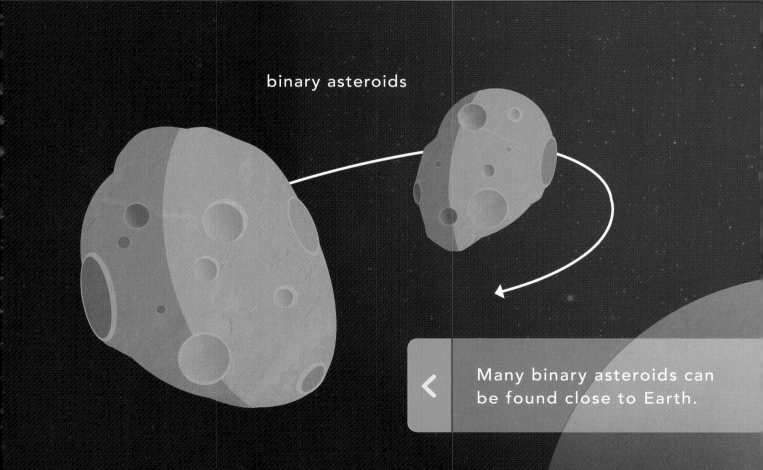

binary asteroids

Many binary asteroids can be found close to Earth.

WHAT DO YOU KNOW?

Tell someone what you know about asteroids! What else can you discover?

⌄

Saturn

→

MAKE POTATO ASTEROIDS

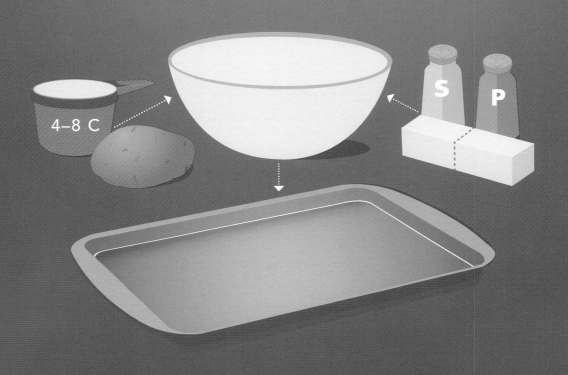

4–8 C

S P

— **What you need** —

4–8 cups mashed potatoes, ½ stick butter, salt and pepper, a greased cookie sheet

What you do

Have an adult heat an oven to 375°. Use ½ cup of mashed potatoes to make each asteroid. Make "craters" by poking your finger into the shape. Place your asteroids on the cookie sheet and bake in the oven for 20 to 25 minutes. Let cool, and then eat your potato asteroids!

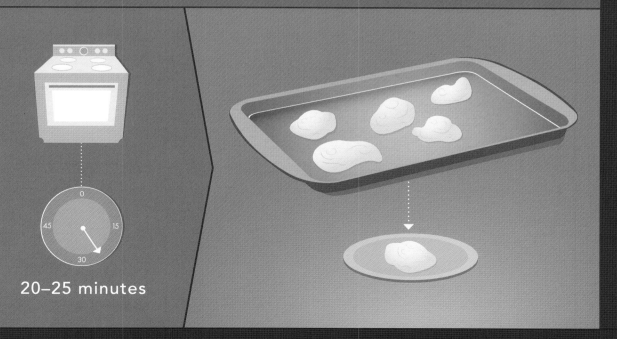

20–25 minutes

GLOSSARY

	binary	something with two parts
	gravity	the force that pulls objects toward each other
	moons	bodies that go around a planet in space
	orbit	the path a planet, moon, or other object takes around something else in outer space
	planets	rounded objects that move around a star
	telescopes	viewing tools that make objects that are far away appear closer

READ MORE

Zobel, Derek. *Asteroids.*
Minneapolis: Bellwether Media, 2010.

Zobel, Derek. *The Hubble Telescope.*
Minneapolis: Bellwether Media, 2010.

WEBSITES

Dawn Mission: Kids
http://dawn.jpl.nasa.gov/DawnKids/
Learn more about asteroids through activities, games, and stories.

Space Janitor
http://kids.nationalgeographic.com/kids/games/actiongames/space-janitor/
Pretend you're an astronaut, and keep space clean!

Note: Every effort has been made to ensure that the websites listed above are suitable for children, that they have educational value, and that they contain no inappropriate material. However, because of the nature of the Internet, it is impossible to guarantee that these sites will remain active indefinitely or that their contents will not be altered.

INDEX

asteroid belt	**4, 6, 15**
astronomers	**4, 8, 12**
binary asteroids	**17**
Ceres	**12, 13**
craters	**5, 21**
Eros	**14, 15**
Gaspra	**15**
Ida	**17**
moons	**13, 17**
near-Earth objects	**11**
Piazzi, Giuseppe	**12**
planets	**4, 6, 13**
spacecraft	**8, 9**
telescopes	**8**